CLASS .... 516 ....

AUTHOR.

BOOK NO. 40066034

CURVES

Leapfrogs

Acknowledgements

The following people were involved in the making of this book; some helped to write it, others were responsible for its design, some were involved with the photographs, the illustrations and the computer plots, while others helped to print it:

Arthur Allen, David Alexander, Anna Baker, Claire Baker, Lyndon Baker, Timothy Blanchard, David Cutting, Linda Davis, Helen Dyal, Ruth Feinstein, Kevin Glennon, Liz Gordon, June Gray, Tansy Hardy, Jeff Hasell, Ray Hemmings, Annamarie Hill, Derick Last, Jack Lewis, David Lowrie, Jenny Luvaglia, Dick Newell, Leo Rogers, David Sturgess, Dick Tahta, Warwick Tomsett, Bob Tonks, Heather Wood, Heidi Yeo, Children and Staff of Crosshall Junior School, Eynesbury, St Neots, and the Staff of Gamlingay Village College and Gamlingay First School, Cambridgeshire.

Written and published by Leapfrogs
Produced by Leapfrogs with EARO

Design and artwork by David Cutting Graphics
Printed by Foister and Jagg Limited, Cambridge

Distributed by Tarquin Publications,
Stradbroke, Diss, Norfolk
ISBN 0 905531 29 9 (hardback)
ISBN 0 905531 30 2 (paperback)

Computer graphics on pages 16—17 by Robary Limited, Cambridge.
Some of the photographs were supplied by the following:
Camera Press Limited, pages 4, 5, 6, 17, 61, 75.
Keystone Press Agency Limited, page 61.
The Natural History Photographic Agency, page 23.
The Victoria & Albert Museum, page 41.

Contents

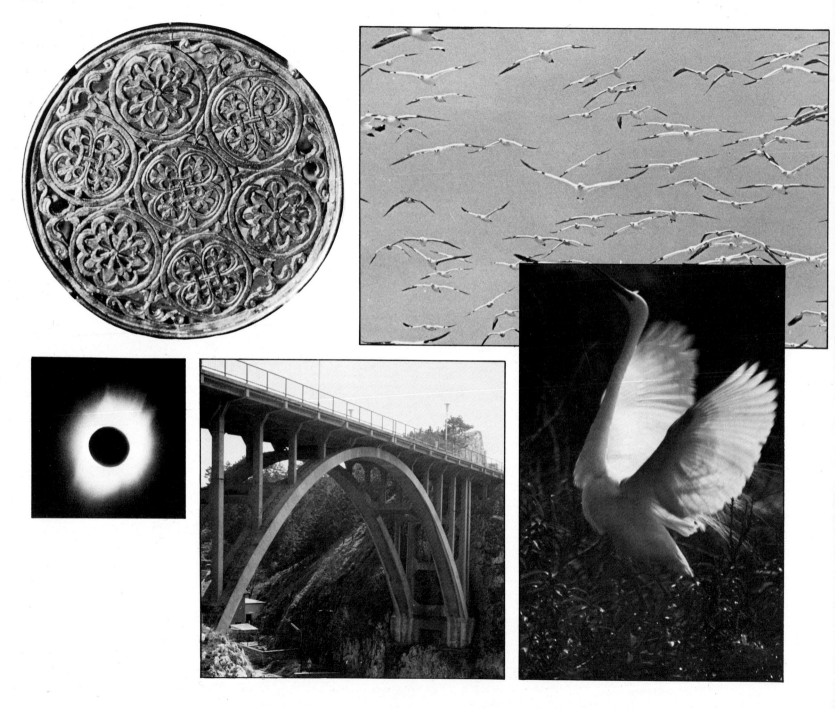

Midpoints

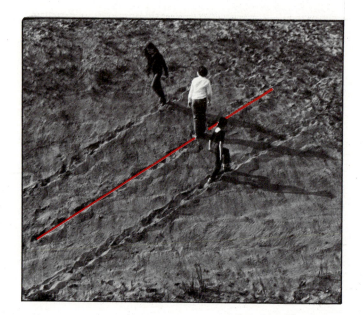

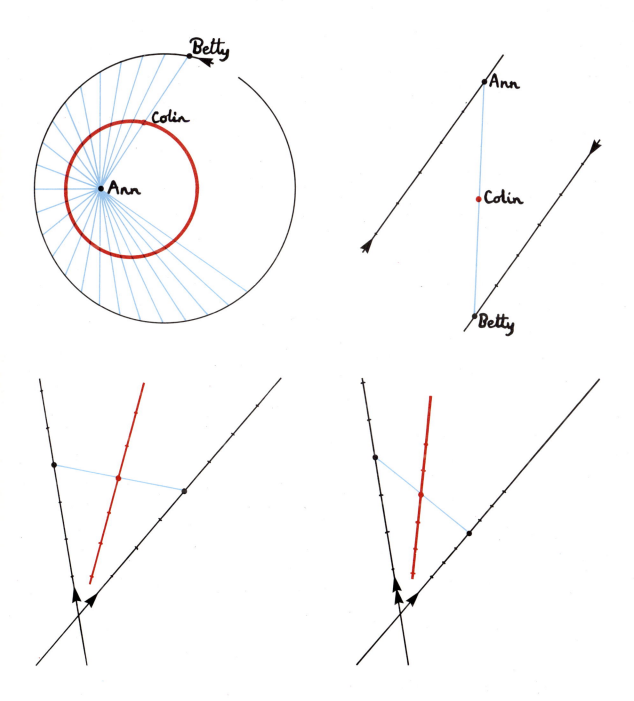

Ann and Betty decide to move according to some rule — by walking in a line or a circle, or perhaps just standing at a point. They have to decide where to start from and whether one of them is going to move faster than the other. In order to move at the same speed they could agree to take one pace together at a time. Otherwise if Ann takes two paces to Betty's one she will be moving at twice Betty's speed.

Speeds are indicated by numbers of arrowheads in the diagrams.
A third person, Colin, joins in by 1) always keeping in a line between Ann and Betty and 2) always keeping halfway between them. The path that Colin has to take in different cases can be found by actually doing this with people or by drawing. The latter can be done accurately by measuring successive positions, or more roughly by guessing the positions of the midpoints by eye. Some simple cases are shown on this page and more complicated ones are explored in the following pages.

Ann moves in a circle, walking at the same speed as Betty who is moving along a line. They start at positions that bring them closest to each other. Colin keeps halfway between them. As Ann and Betty move they will in fact get further and further apart. Colin may be surprised to find that every time Ann completes a circle he repeats a certain movement. After walking half her circle, Ann suddenly seems to be moving away from Betty and at this point Colin has to change direction sharply, producing a sort of kink in his curve that is called a *cusp*.

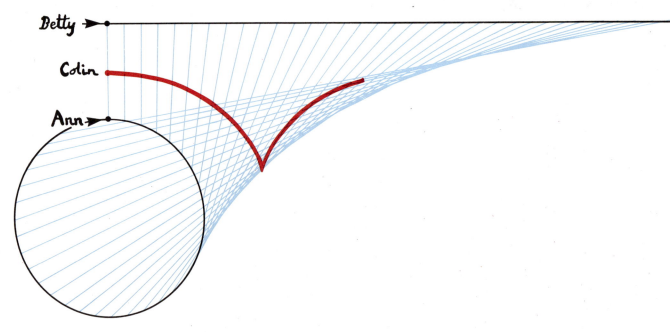

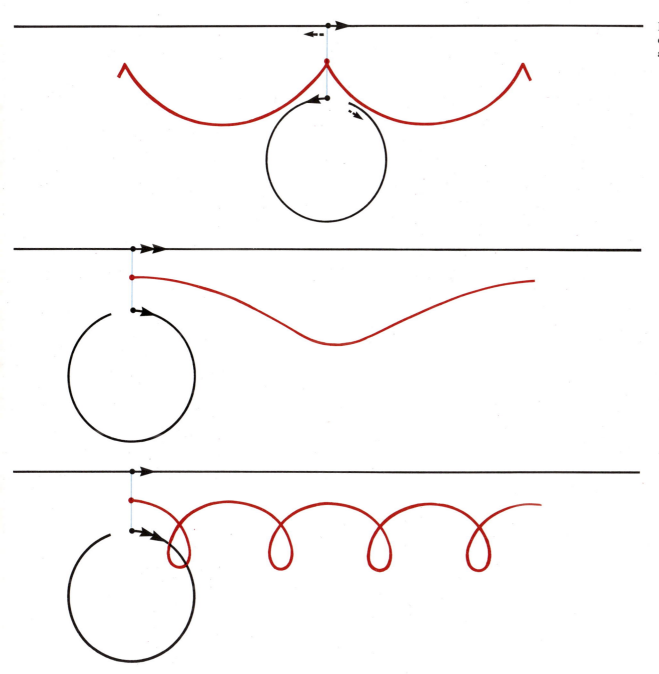

Diagrams illustrate the different cases when Ann or Betty move at different speeds.

Ann moves in a circle, Betty on a line touching the circle at one point. They start at the point where the tangent touches the circle and move at the same speed. Colin keeps halfway between them. Colin's path turns out to be a curve known as a *cycloid*. This path is the one traced by a point on the rim of a rolling wheel as seen at the foot of the picture opposite.

The diagrams show the fact that the two different methods of construction yield the same curve. Two different sets of positions A, B, C and A′, B′, C′ for Ann, Betty and Colin are shown. The midpoints C,C′ and K (the midpoint of OB) are such that C lies on a circle, centre K, and radius one half that of Ann's circle. This circle, centre K, rolls on the line OC′.

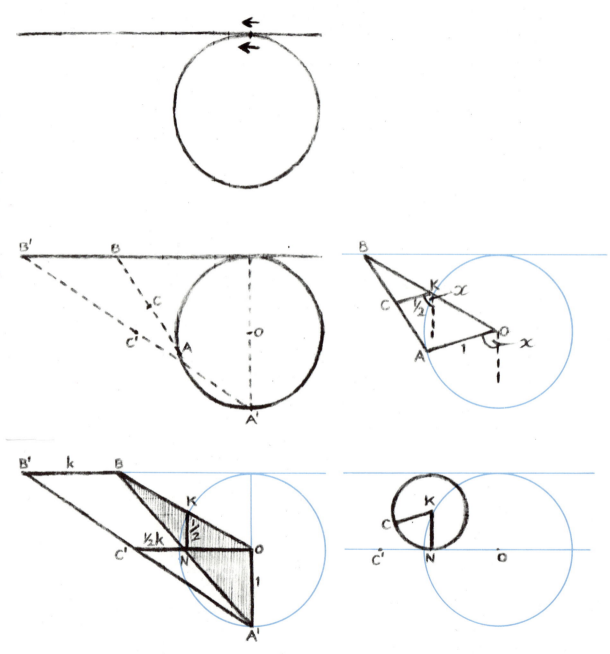

The cycloid produces an attractive arch that has sometimes been used in building bridges.

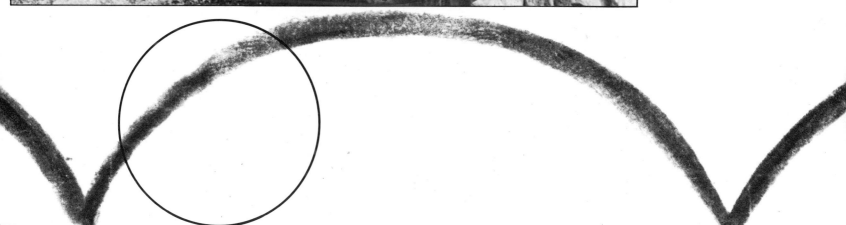

Ann and Betty both move on circles and Colin keeps halfway between them. There are an enormous number of different possibilities, depending on where Ann and Betty start and at what speeds and in what direction they move. But it does not matter how far from each other the circles are, for it turns out that Colin's path will always be *as if* Ann and Betty were moving on circles with centres at the same point.

This remarkable result is indicated by diagrams. When Ann and Betty are moving on two circles, centres X and Y, Colin moves in exactly the same path as if Ann and Betty were moving on the same sized circles but both with centre at Z, the midpoint of X and Y. So in the following pages the circles will always be taken to have their centres at the same point. Colin's path in various cases may be explored actually using people, or by drawing as suggested opposite. Ann's and Betty's movements may be thought of as those of the hands of a clock, rotating with different speeds and in the same or opposite directions.

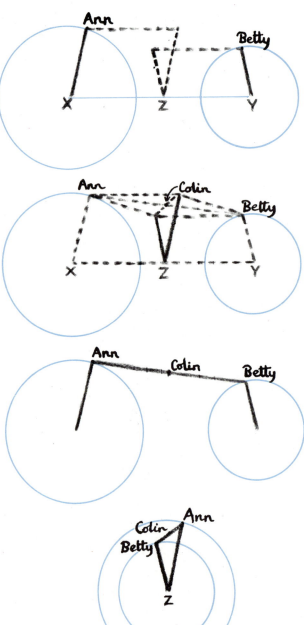

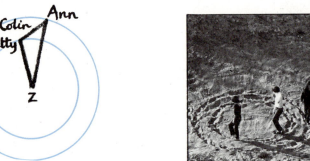

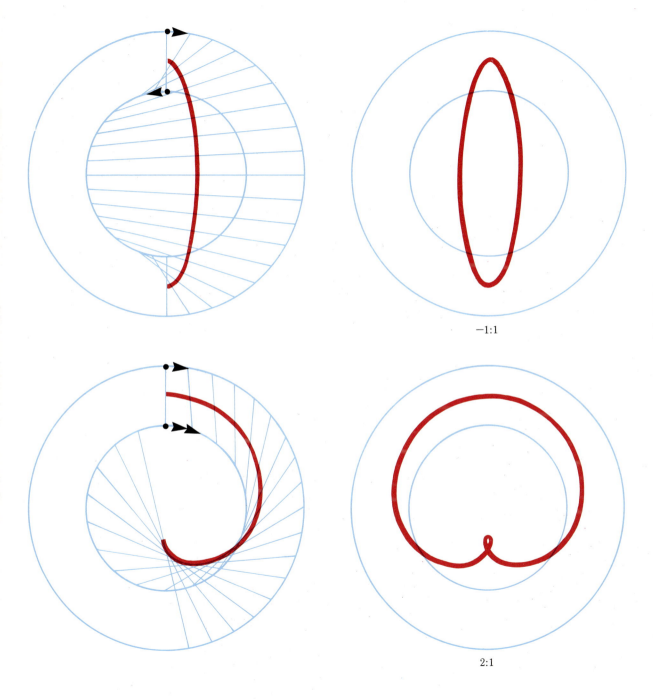

−1:1

2:1

In the first case the circles are traced in opposite directions at the same speed. This is indicated by the ratio −1:1.

In the second case the circles are traced in the same direction, the smaller one turning at twice the speed of the other. This is indicated by double arrowheads in the diagram and by the ratio 2:1.

Note that the notation refers to rates of turning (*angular velocities*) and not to the actual speeds of Ann and Betty who are here moving on different circles. Thus, in the first case, Ann is in fact moving faster than Betty though her radius is turning at the same rate as Betty's.

In each case the hands start in the same position, pointing upwards.

These show the resulting paths for various different cases. The curves arising when the circles are traced in the same direction are called *epitrochoids*, those arising when the circles are traced in opposite directions being called *hypotrochoids*.

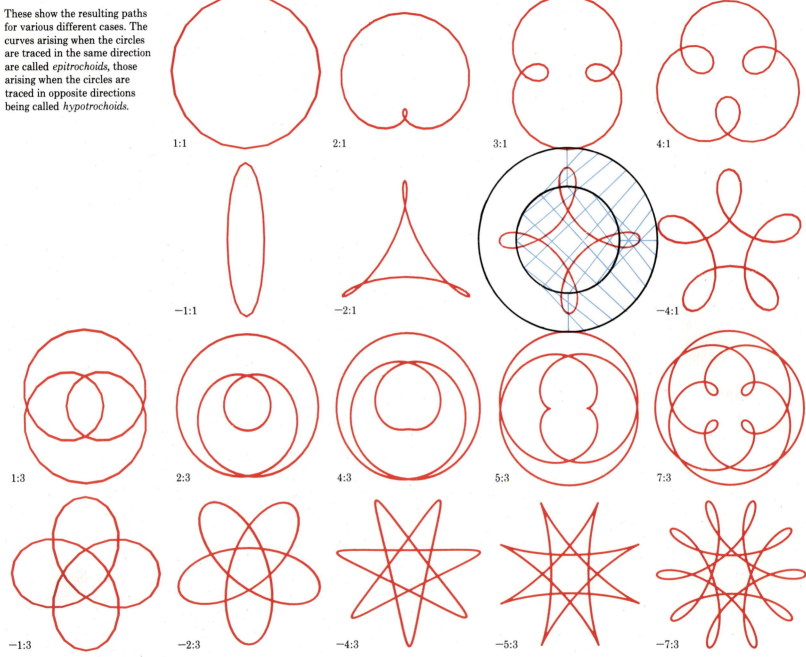

1:1 2:1 3:1 4:1

−1:1 −2:1 −4:1

1:3 2:3 4:3 5:3 7:3

−1:3 −2:3 −4:3 −5:3 −7:3

4:3

7:2

−4:3

−7:2

The paths arise from particular cases where the circles are the same size. Thus Ann and Betty are here moving round the same circle — at different speeds and sometimes in opposite directions — having started from the same position. The resulting curves are sometimes called *rose-curves* or *folia* (eg *trifolium, quadrifolium*, etc).

Some particular cases arise
when the circles are traced at
speeds that are in the same ratio
as the lengths of each other's
radii. For example, when one is
three times larger than the other
but is traced at one third the
speed of the other. In such cases
the motions yield paths that
could also be generated by points
on a circle rolling, inside or
outside, another fixed circle.
Such paths are usually called
epicycloids (when the circle rolls
outside the fixed one) and
hypocycloids (when the circle
rolls inside).
The diagrams indicate the
equivalence of the two
definitions.
Some special cases are shown on
the following pages and these
have some further special
names.

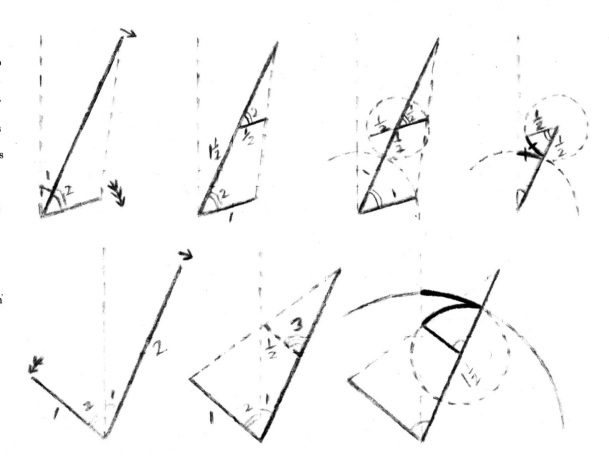

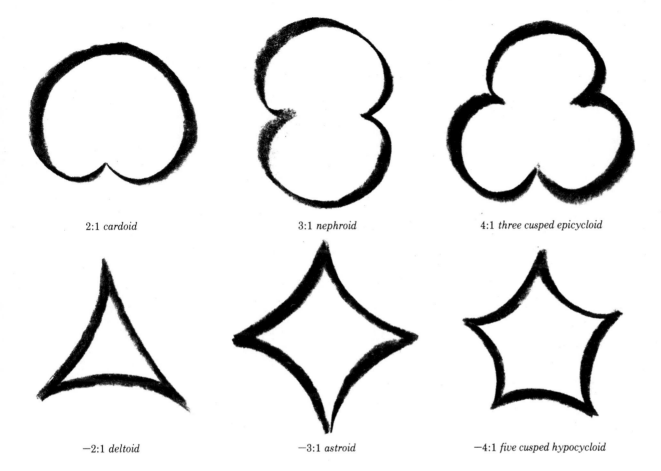

2:1 *cardoid*   3:1 *nephroid*   4:1 *three cusped epicycloid*

−2:1 *deltoid*   −3:1 *astroid*   −4:1 *five cusped hypocycloid*

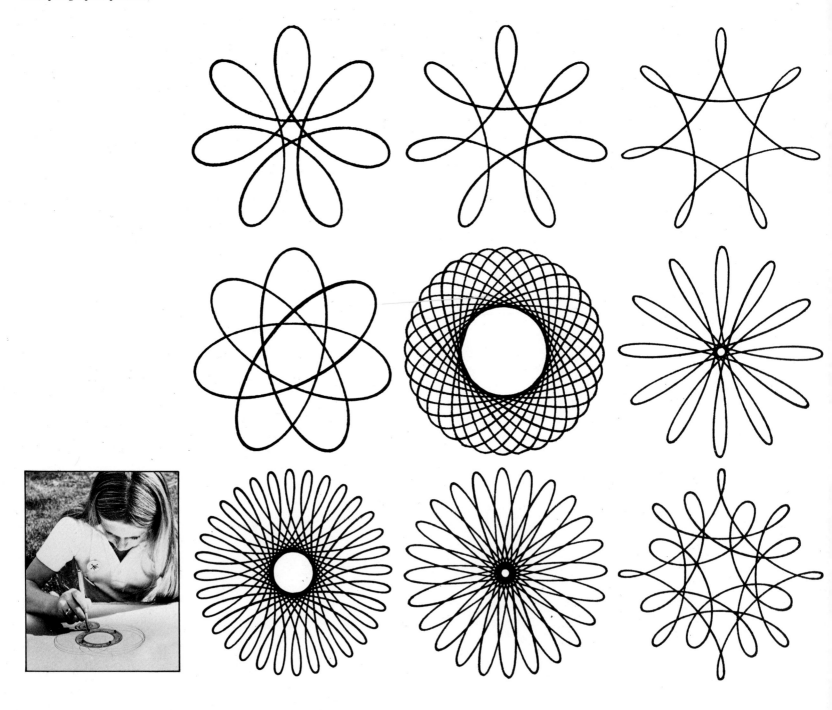

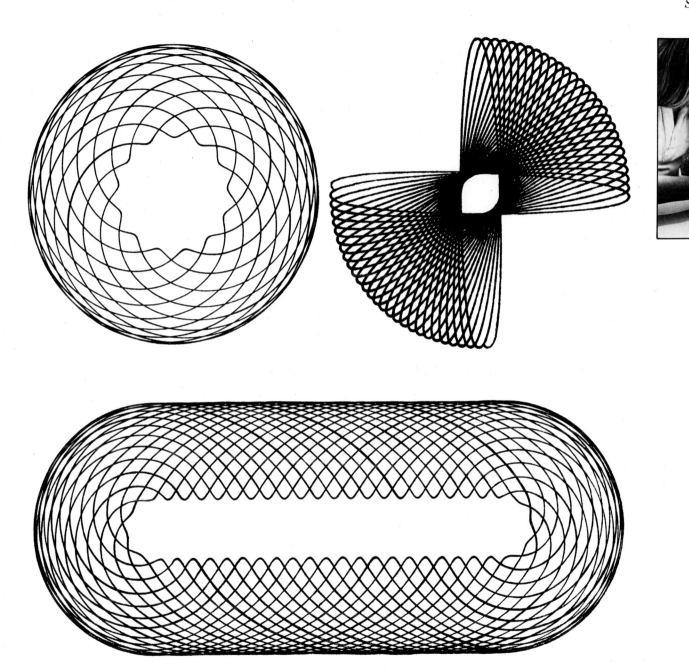

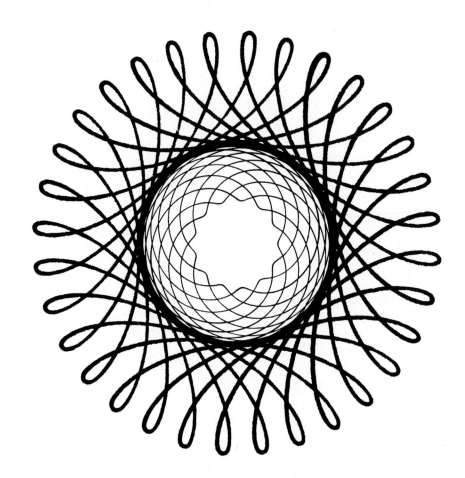

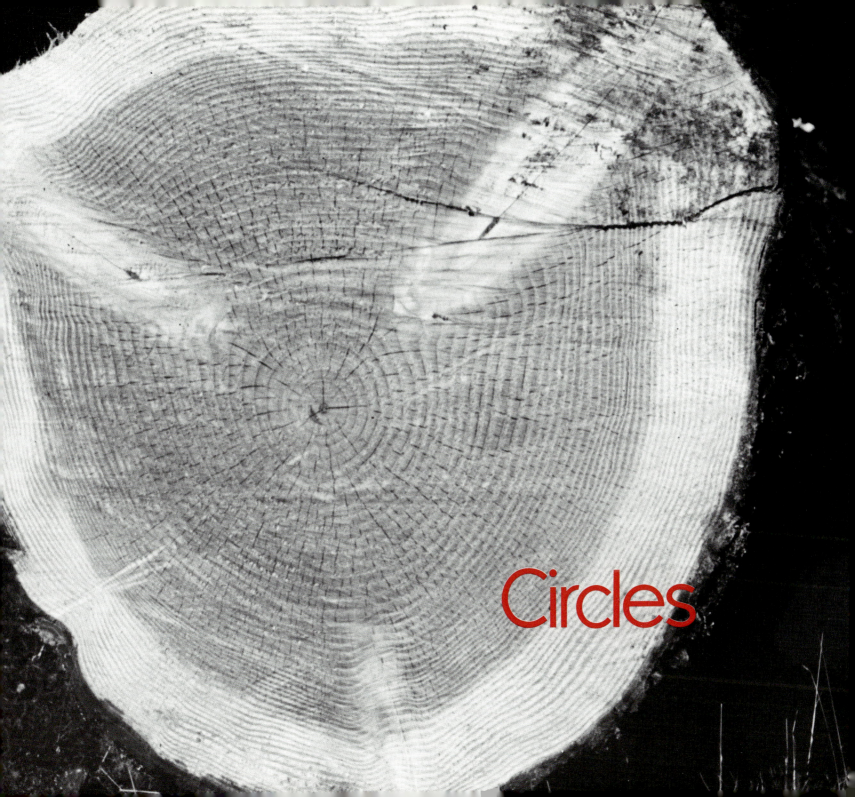

Circles

Families of circles can be drawn using compasses to produce various interesting designs. Sometimes these yield an outline that suggests some curve, called the *envelope* of the family. One of the simplest methods of drawing a family is to choose a base-circle and a base-point. Then a family of circles is drawn each having its centre on the circumference of the base circle and a radius such that it passes through the base-point. Some cases are shown with the base-point inside, on and outside, the base-circle. The resulting envelopes are called *limaçons*, the special case when the base-point is on the base-circle being called a *cardoid*.

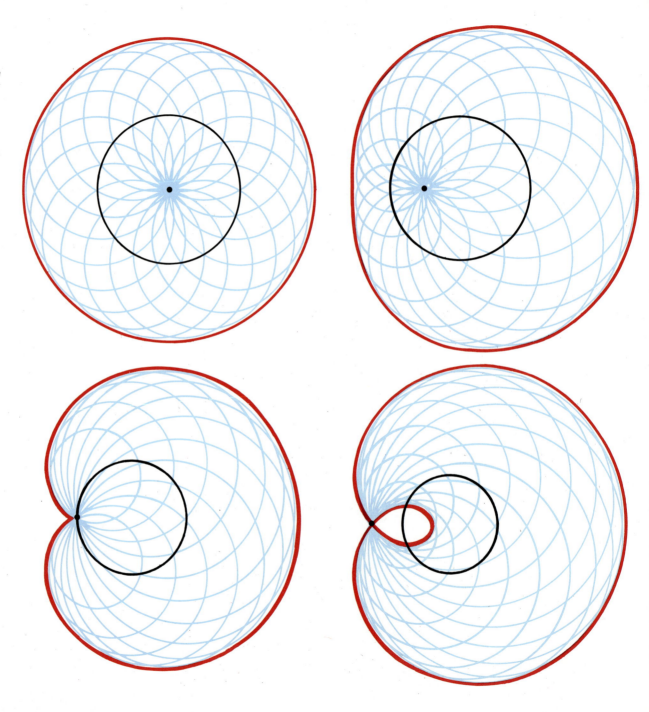

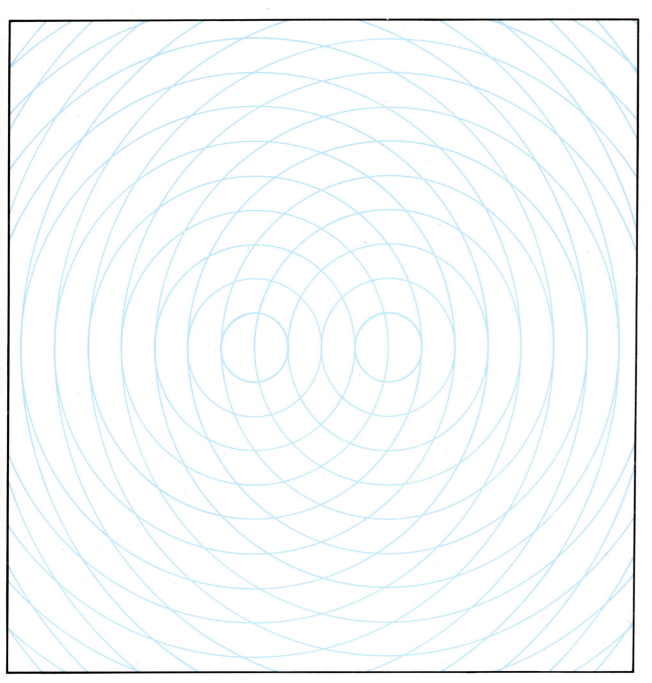

Families of concentric circles are like ripples in a pond. They may be thought of as growing in the same way. More specifically a set of concentric circles may be thought of as being the positions of one circle that is growing or shrinking at various speeds.

Two families of concentric
circles will intersect in lots of
points. We can look at particular
points by taking circles from
each family that correspond in
some way. For instance, if we
think of each family as being
generated by one circle growing
and shrinking in some way and
perhaps intersecting at each
moment the corresponding circle
that is tracing the other family.
The first example shows two
families being traced at the
same speed and from the same
size starting point. The second
shows what happens when one
family is being traced at twice
the speed of the other, this being
indicated by the ratio 1:2.

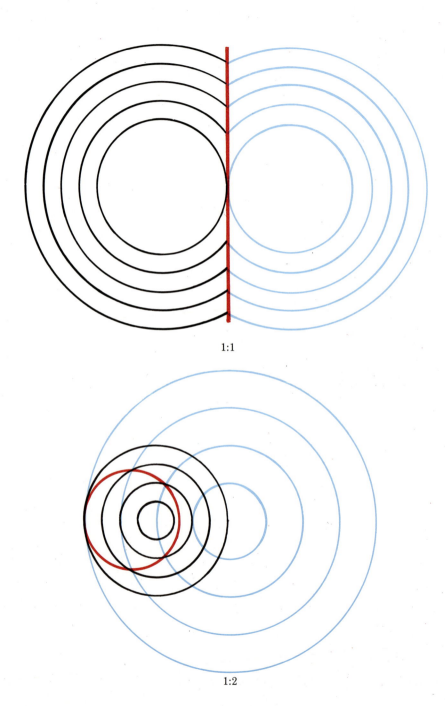

1:1

1:2

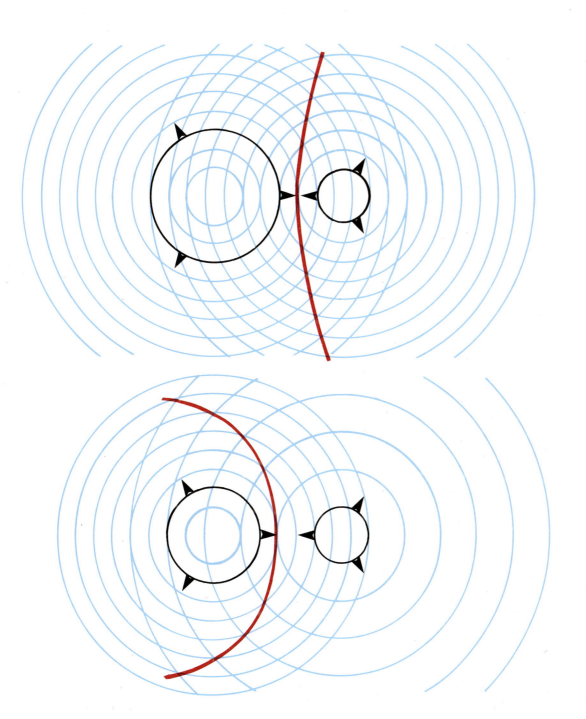

Other examples show what
happens when the growing and
shrinking arises from *staggered*
starts — as if ripples in a pond
were produced by stones thrown
in at different times.

1:1 (staggered)

1:2 (staggered)

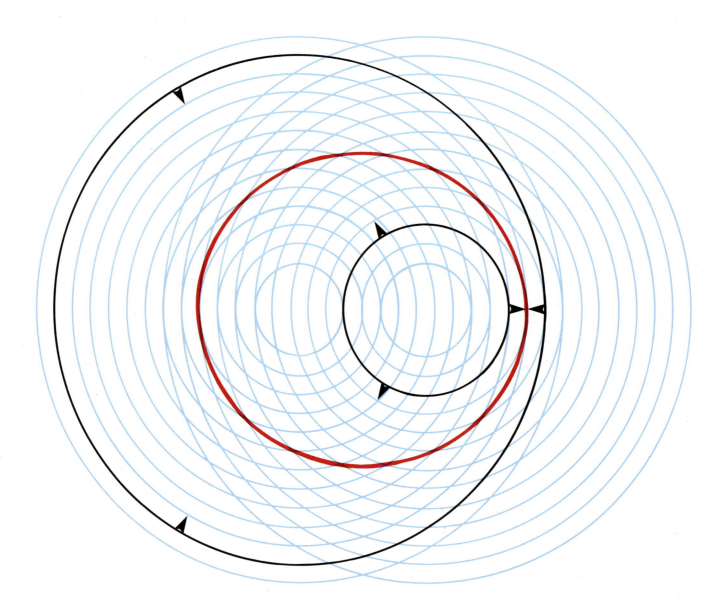

−1:1 (staggered)

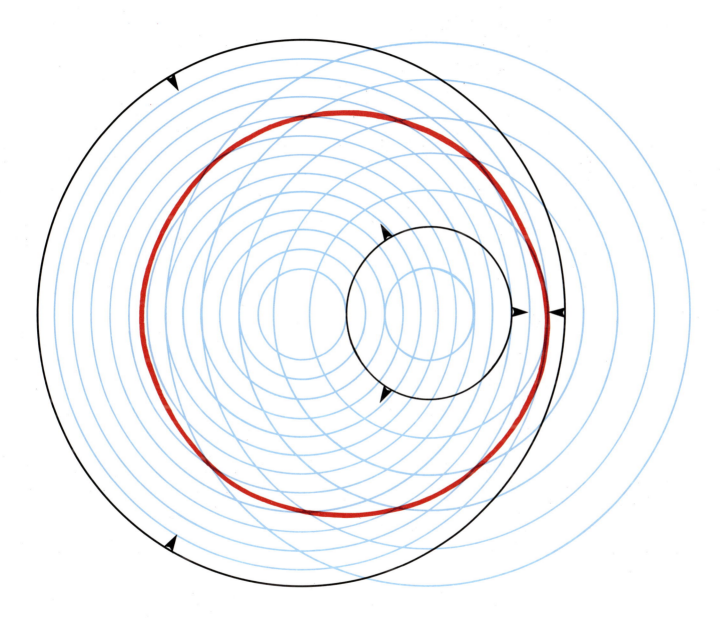

−1:2 (staggered)

The sun and the moon *appear* to be about the same size, although the sun is really about 400 times as big as the moon. Of course, the sun is much further away, so that the *viewing angle* for each body is about the same — in fact about half a degree.

Two unequal diameters can appear to have the same length. In the same way, two equal lines could appear to have different lengths. It all depends where they are viewed from. Where are the points from which two equal lines will in fact also *appear* equal?

Such points will have equal viewing angles. They can be constructed as intersections of two equal-sized circles, each passing through the endpoints of one of the lines. However, some intersections could be such that the point is on the smaller part of one circle but on the larger part of the other. In such a case the viewing angles are not equal, they are in fact supplementary, that is they add up to 180 degrees and the lines will not appear equal.

The diagrams show the positions of points with equal viewing angles in various cases. They also indicate the positions of points with supplementary viewing angles. Finally the combined positions — obtained by taking all possible intersections of pairs of equal circles — are shown from a different point of view.

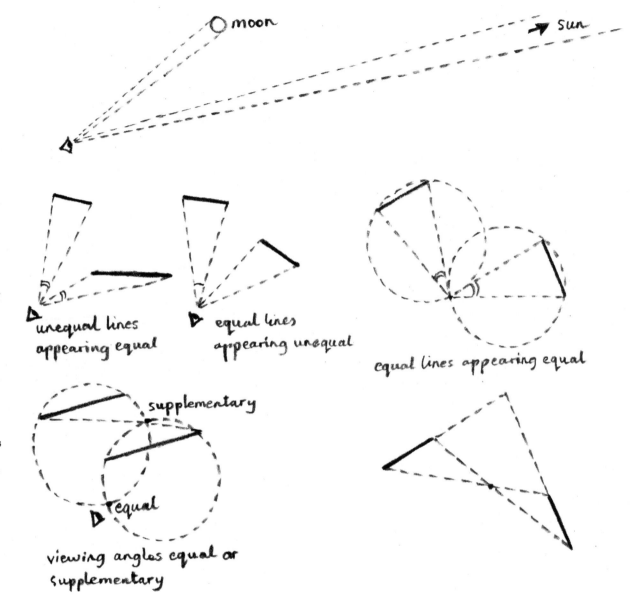

moon

sun

unequal lines appearing equal

equal lines appearing unequal

equal lines appearing equal

supplementary

equal

viewing angles equal or supplementary

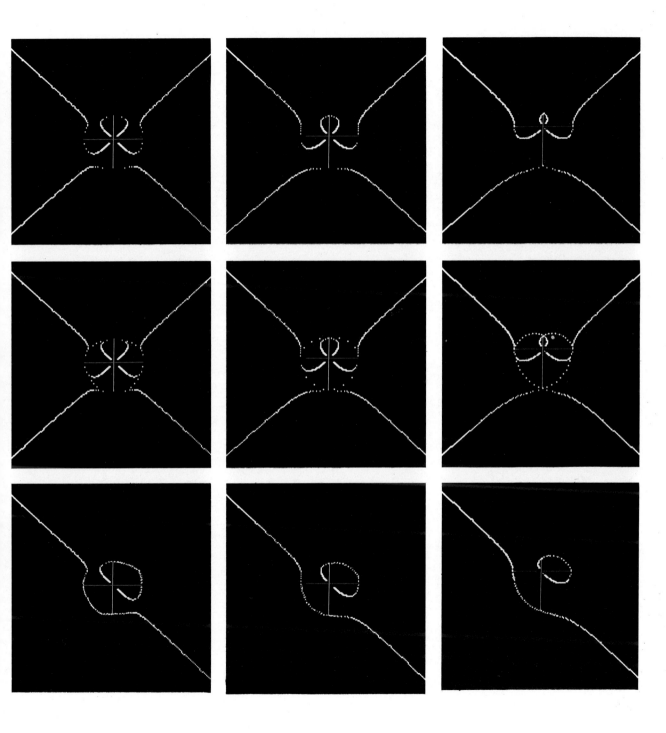

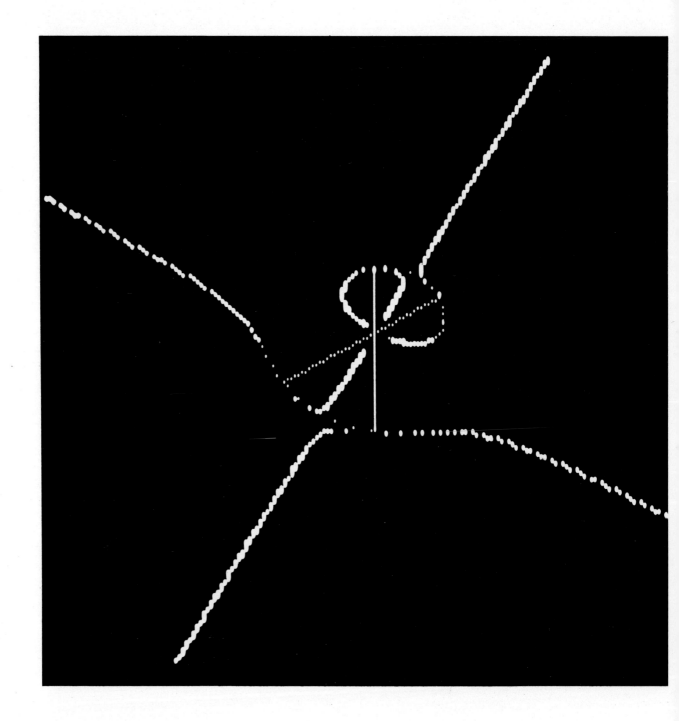

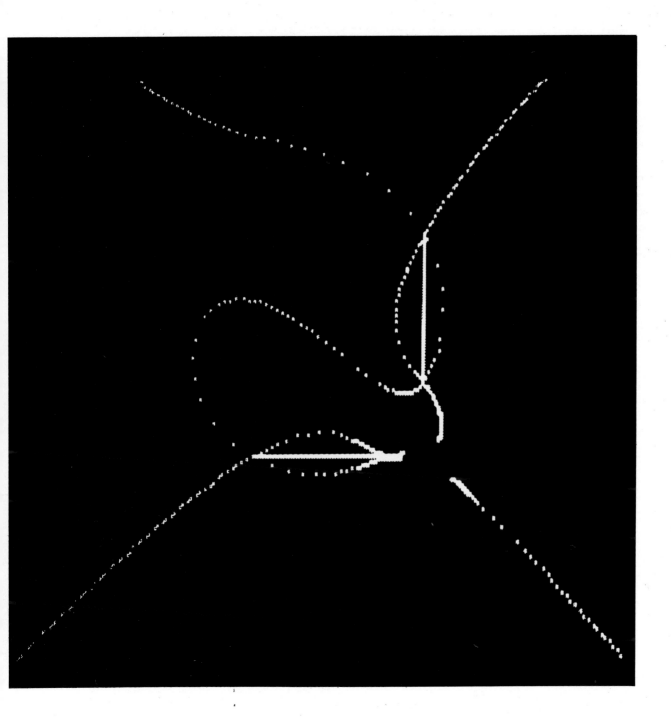

A rotating line may be thought of as generating a *star*. The line and its various positions are taken to extend outwards without limit; they are infinite. Two stars of this sort produce lots of intersections. We can look at particular intersections by taking lines from each star that correspond in some way. This may be done by thinking of the rotating line intersecting the corresponding line that is tracing the other star.

For example, one line may be turning clockwise — starting, say, from a horizontal position — and the other may be turning in the same direction but at twice the speed. This is indicated by the ratio 1:2. Various other cases are shown.

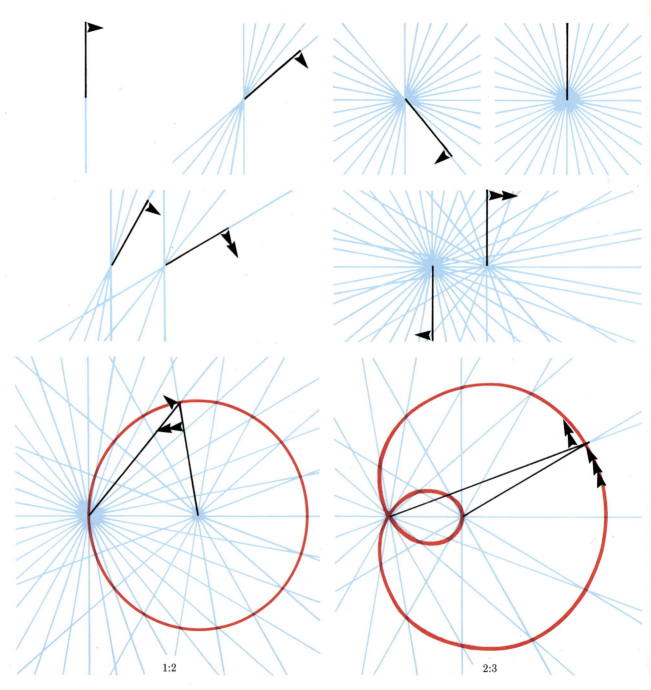

1:2

2:3

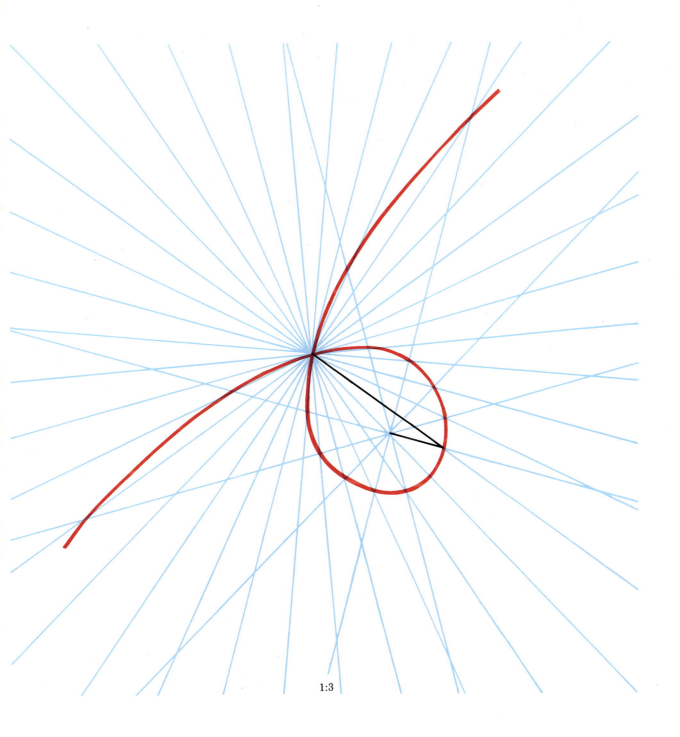

1:3

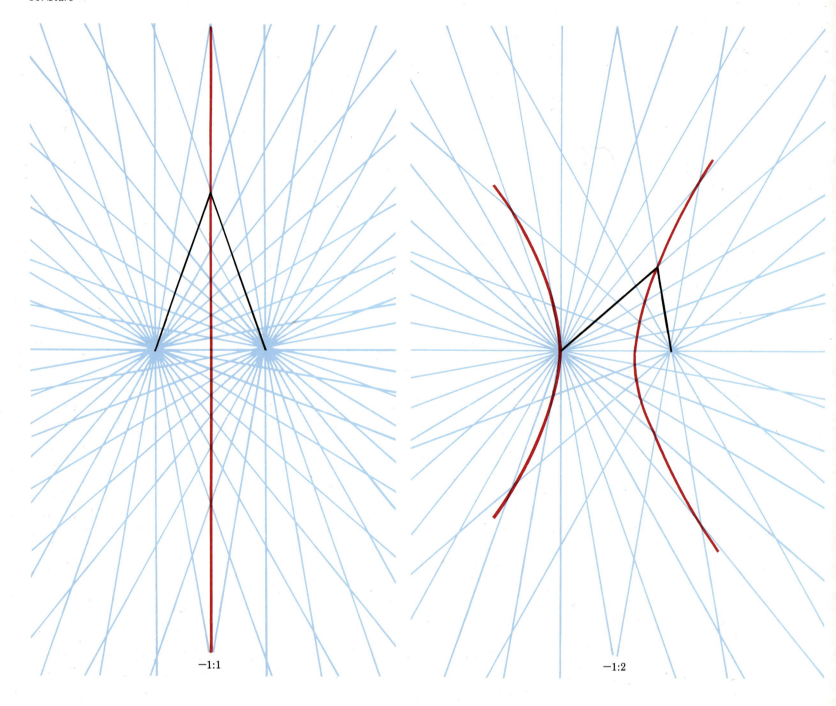

−1:1

−1:2

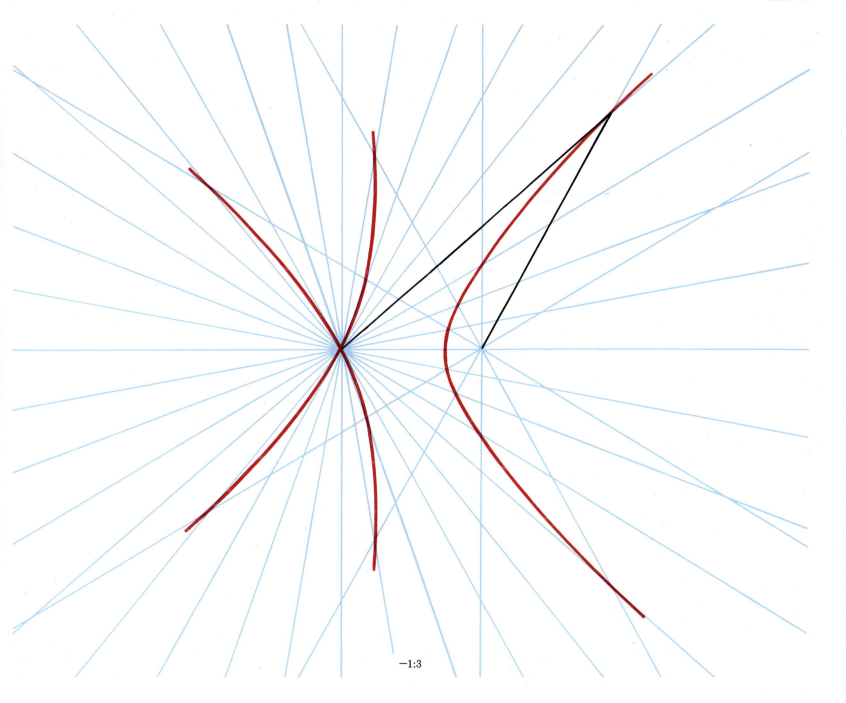

−1:3

Ann stands still and Betty moves in a circle round her. Colin always keeps in a line between the girls but keeps moving outwards from Ann at a constant speed.

The motion could be reconstructed by placing a circular piece of card on the turntable of a record player and drawing a line from the centre. Otherwise the path could be marked on a *polar grid*, the points move out the same distance along successive spokes of the grid.

The resulting curve could be called an *equidistant spiral* though it is usually known as an *Archimedes spiral*. The fixed distance stepped out each time round the polar grid is why this is the shape taken up by a coil of rope.

Some amusing variations of the construction are shown.

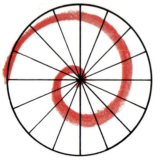

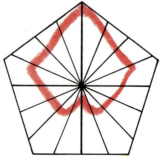

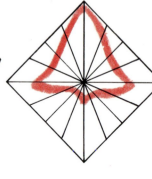

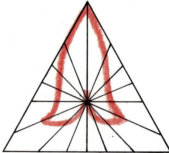

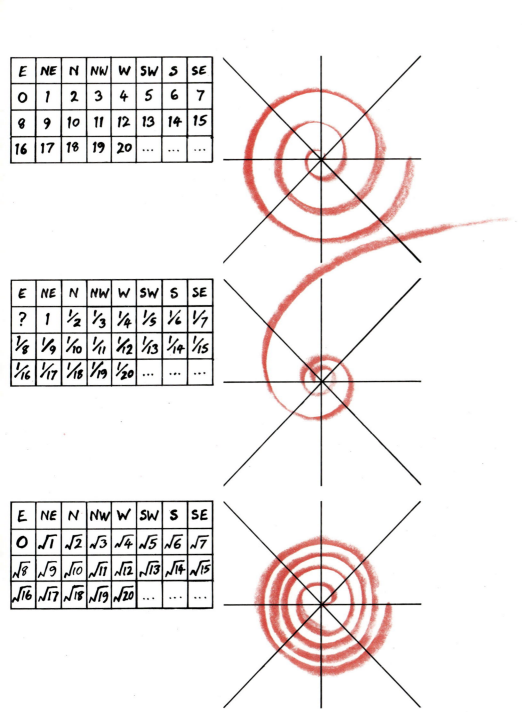

E	NE	N	NW	W	SW	S	SE
0	1	2	3	4	5	6	7
8	9	10	11	12	13	14	15
16	17	18	19	20	...	...	...

E	NE	N	NW	W	SW	S	SE
?	1	$\frac{1}{2}$	$\frac{1}{3}$	$\frac{1}{4}$	$\frac{1}{5}$	$\frac{1}{6}$	$\frac{1}{7}$
$\frac{1}{8}$	$\frac{1}{9}$	$\frac{1}{10}$	$\frac{1}{11}$	$\frac{1}{12}$	$\frac{1}{13}$	$\frac{1}{14}$	$\frac{1}{15}$
$\frac{1}{16}$	$\frac{1}{17}$	$\frac{1}{18}$	$\frac{1}{19}$	$\frac{1}{20}$	...	...	...

E	NE	N	NW	W	SW	S	SE
0	$\sqrt{1}$	$\sqrt{2}$	$\sqrt{3}$	$\sqrt{4}$	$\sqrt{5}$	$\sqrt{6}$	$\sqrt{7}$
$\sqrt{8}$	$\sqrt{9}$	$\sqrt{10}$	$\sqrt{11}$	$\sqrt{12}$	$\sqrt{13}$	$\sqrt{14}$	$\sqrt{15}$
$\sqrt{16}$	$\sqrt{17}$	$\sqrt{18}$	$\sqrt{19}$	$\sqrt{20}$	...	...	...

Other ways of plotting points of a spiral on a polar grid are shown. In these cases, some number sequence is chosen and allocated to the eight spokes of the compass. The appropriate distances are calculated and plotted at each spoke.

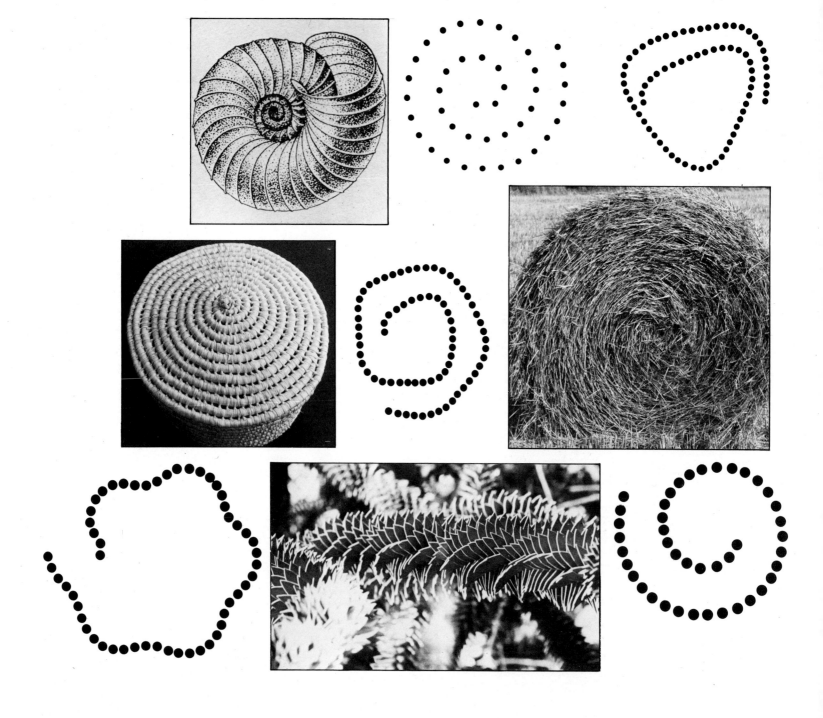

Lines

Ann and Betty move along certain paths. This time they consider the way the line joining them twists and turns as they move. Sometimes the positions of these lines implies the outline of some curve, the *envelope* of the lines. The situation is familiar in the form of so-called *curve-stitching*. This may be thought of as yielding an envelope as Ann and Betty move along lines in opposite directions at the same speed. The envelope would not be clear when this is done with people. It is best shown by drawing or actual stitching. The usual way of doing this produces a curve which is a *parabola*.

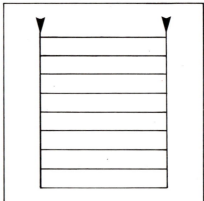

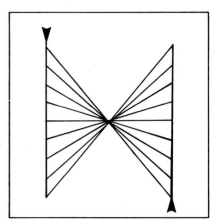

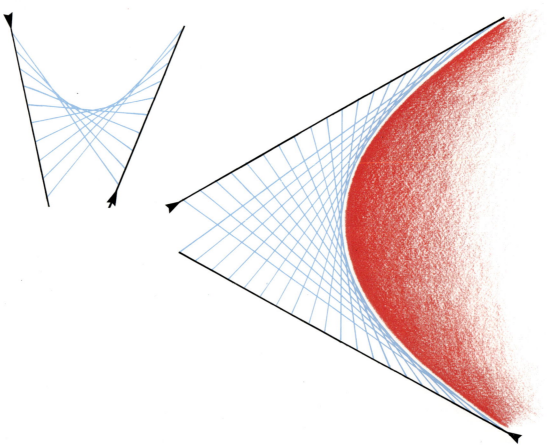

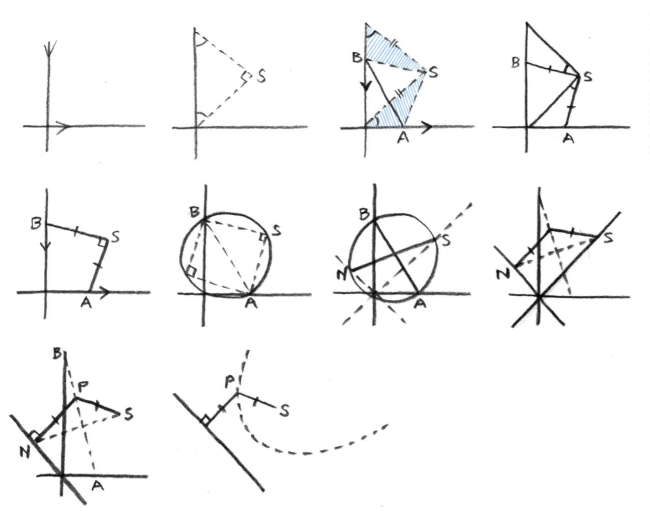

The diagrams indicate why curve-stitching yields a parabola in the sense that the construction yields a curve whose points P are such that they are all at the same distance from a point S, called the *focus*, and a certain line, called the *directrix*. The lines of the curve-stitching construction are tangents to the parabola at the points P.

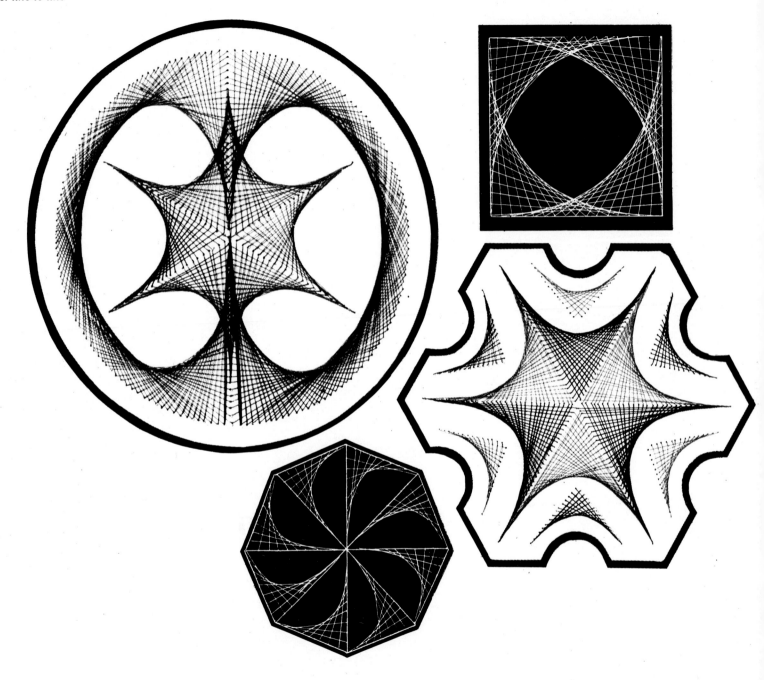

If a parabola can be obtained by curve-stitching, could a similar method yield the other conics — *ellipses* or *hyperbolas*?
This can be done by choosing the movements of the points (Ann and Betty) along the base-lines in a special way. In the case of the parabola the points move in such a way that each is at the same distance from its starting point as the other is. When the movement is such that the product of their distances from two fixed points is always the same then other curves arise.
Some particular cases for parallel and intersecting base-lines are shown.

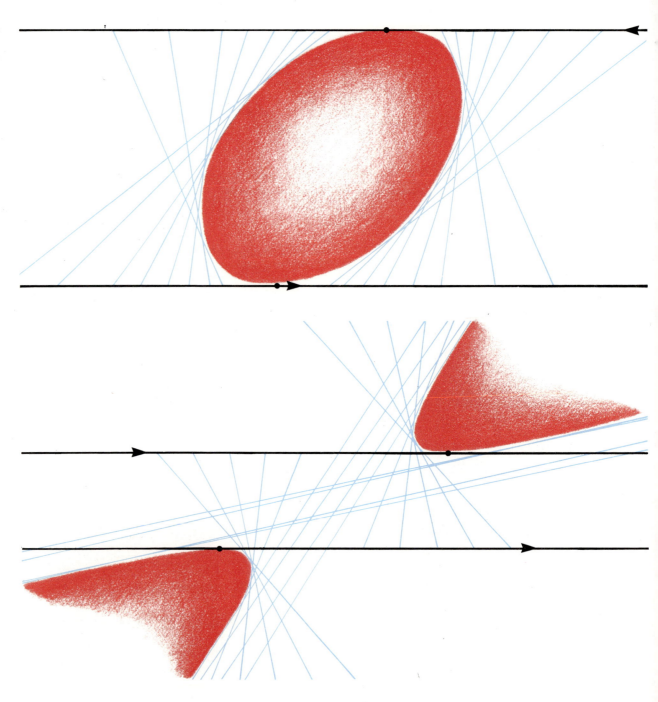

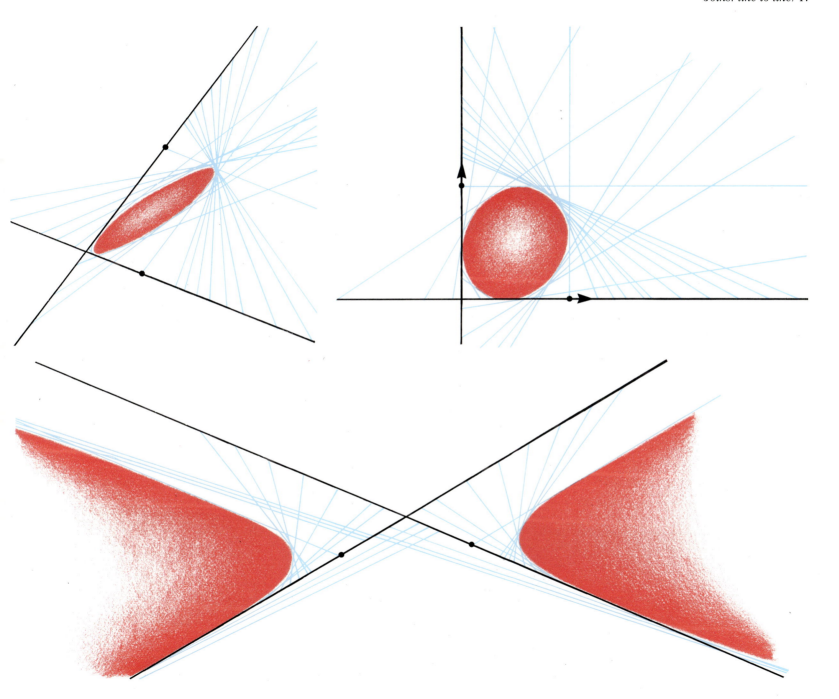

The previous examples of curve-stitching may be thought of as line-to-line stitching. In this case Ann and Betty move round the same circle in some way. This can yield some interesting circle stitchings. Since Ann and Betty are moving on the same circle, their motion may be compared to that of equal clock hands and it is worth comparing the envelope of the line joining the ends of the hands with the path, or *locus*, of the midpoints of the ends (for the latter see p11).

In the following diagrams relative speeds are indicated by arrowheads and opposite directions of motion by a negative sign. The curves shown are all particular cases of epi- or hypo-cycloids.

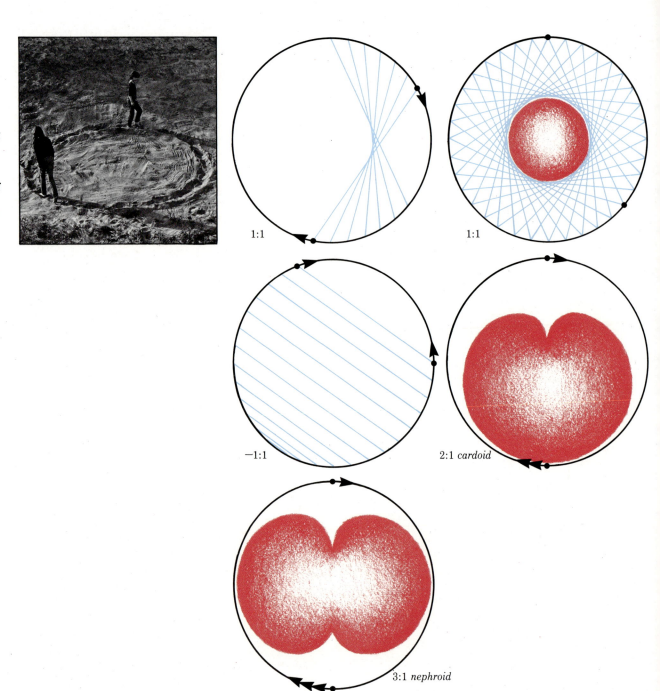

1:1

1:1

−1:1

2:1 *cardoid*

3:1 *nephroid*

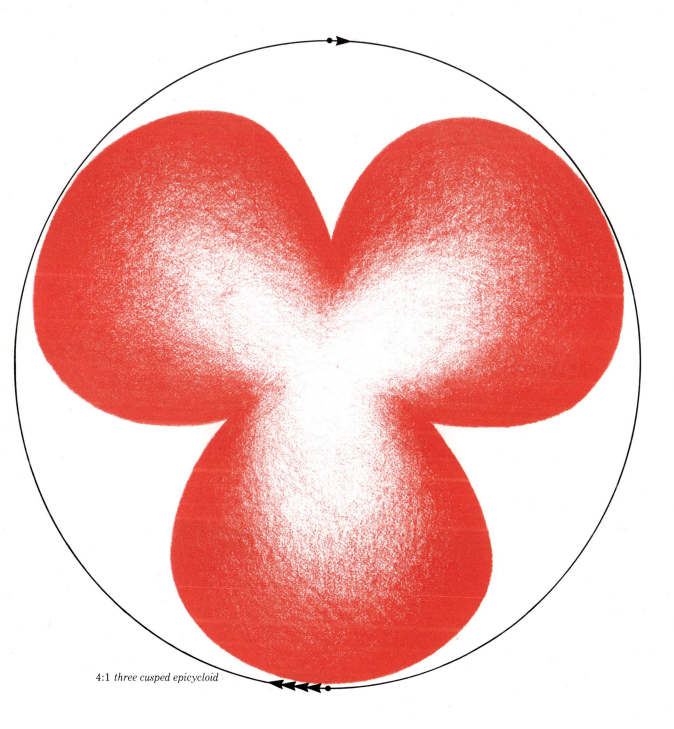

4:1 *three cusped epicycloid*

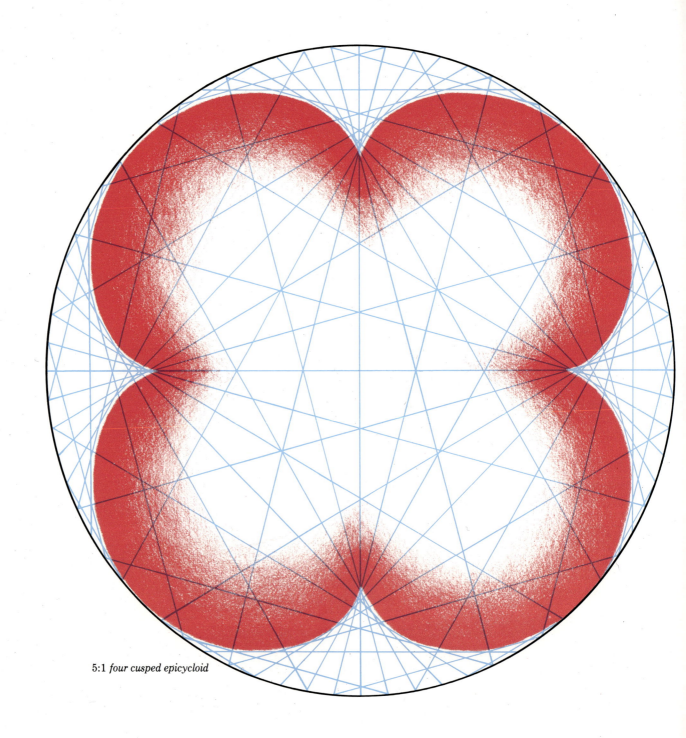

5:1 *four cusped epicycloid*

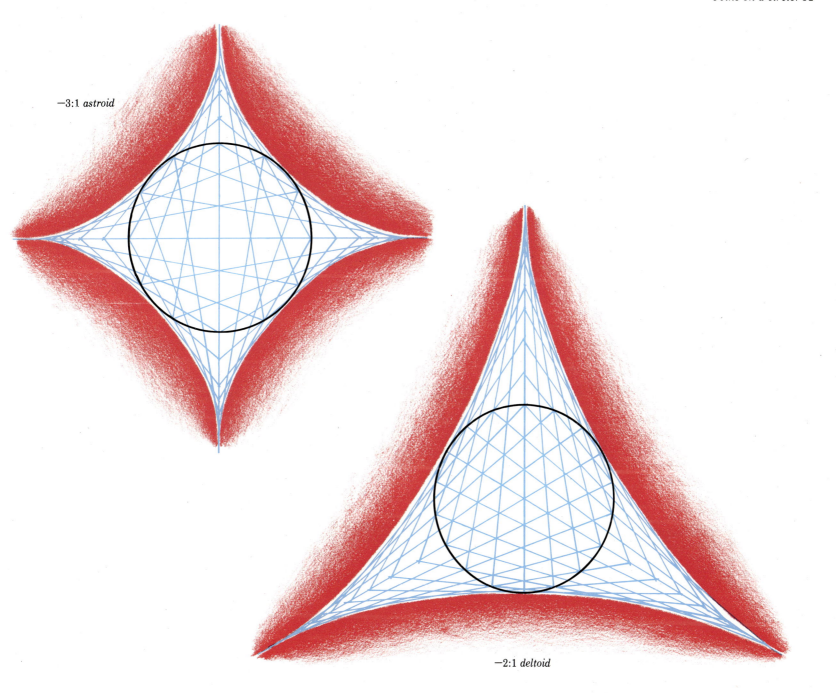

−3:1 *astroid*

−2:1 *deltoid*

Ann and Betty are moving along straight lines, but this time they are holding a long pole or a fully stretched length of rope between them. They are thus only allowed to move in such a way that they remain the same distance apart at all times. Colin moves so as to keep at the midpoint of this distance. His path may be explored by using people, by making drawings using a marked section of a ruler for the fixed distance, or — as shown — by using an acetate sheet sliding over a piece of paper.

(In the latter case, mark the intersecting lines on the paper and the fixed distance on the acetate sheet. Slide the acetate sheet so that the lines pass through the endpoints of the fixed distance and mark successive positions of the midpoint on the acetate. Note also what happens when the roles of paper and acetate are reversed.)

Here, Colin's path is in general an *ellipse*. When Ann and Betty move on some polygonal paths, Colin's path will consist of portions of various ellipses that taken together can yield some amusing shapes.

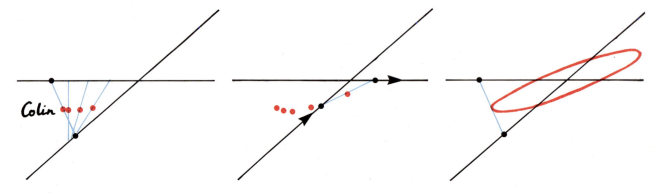

Colin

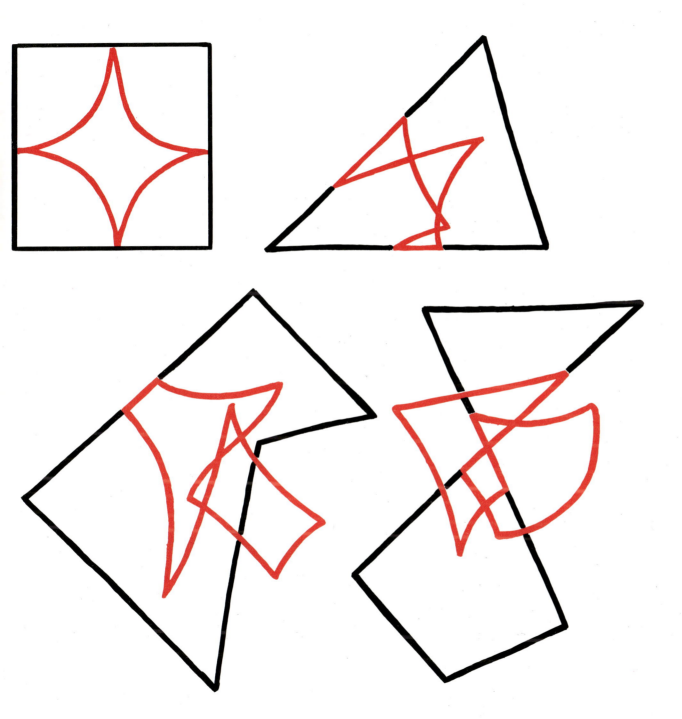

In the case when Ann and Betty
move on perpendicular lines
keeping at a fixed distance from
each other then Colin, moving
with the midpoint, moves in a
circle. If Colin moves with some
other point of the line joining
Ann and Betty, his path will be
an ellipse. That this is so is
indicated by the diagrams which
show that PN is a fixed
'foreshortening' of the distance
QN. This means that the locus
of P is a *shear* of a circle and
this is a convenient and stand-
ard definition of an ellipse.
This construction is the basis of
a mechanical method of drawing
ellipses known as a *trammel*. A
strip has two ends moving in
perpendicular grooves and a pen
can then trace the curve through
a hole in the strip.

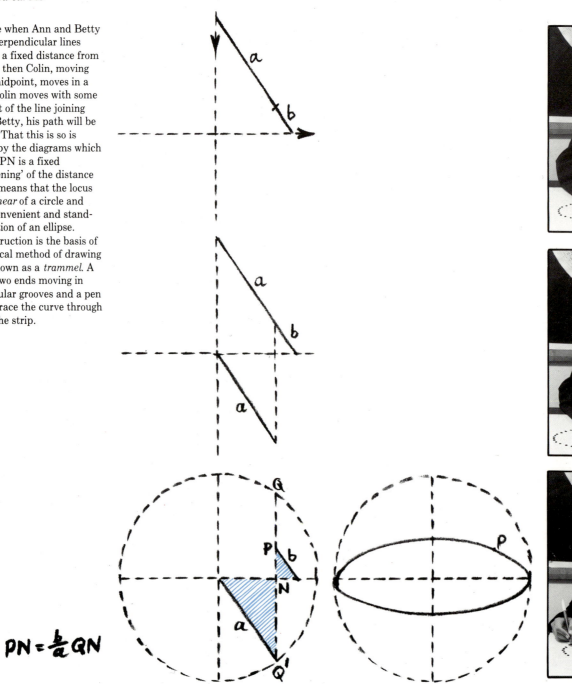

$$PN = \frac{b}{a} QN$$

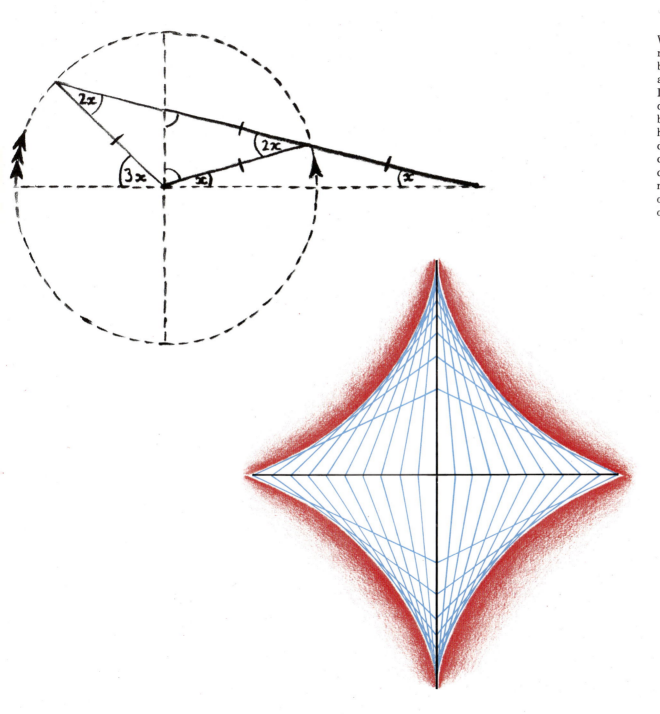

When Ann and Betty are moving along lines holding a pole between them, this pole twists and turns in an interesting way. Its envelope is best found by drawing. That this turns out to be an *astroid* (or four-cusped hypocycloid) is indicated by the diagram which shows that the construction is equivalent to circle-stitching between points moving in opposite directions, one three times the speed of the other (cf p51).

This time Ann is moving in a circle while Betty moves in a line but they are still keeping the same distance apart all the time. The motion could be reconstructed by using two strips joined flexibly with one free end being fixed and the other made to keep on a fixed line. This *two-bar linkage* could be made using plastic geostrips or homemade strips cut from card. Otherwise the motion could be reconstructed using a fixed distance on ruler or card and sliding this between a circle and a line.

Some particular cases are shown indicating how Colin moves with the midpoint.

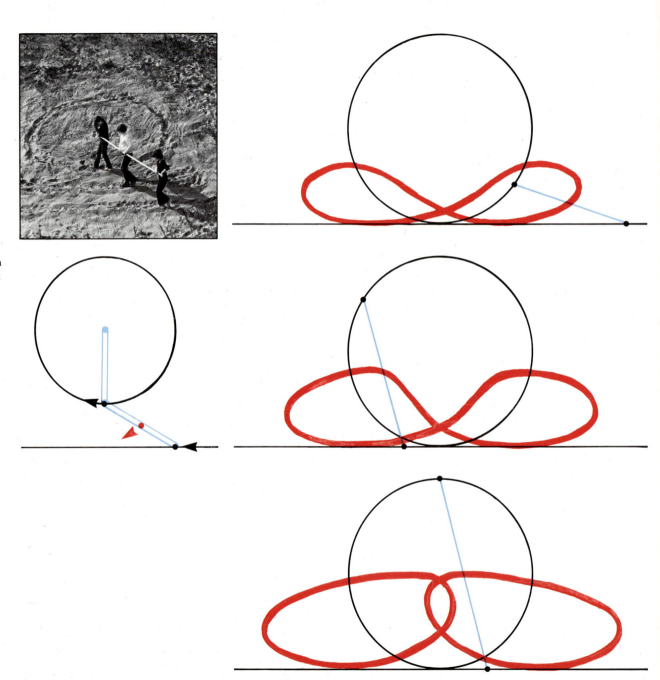

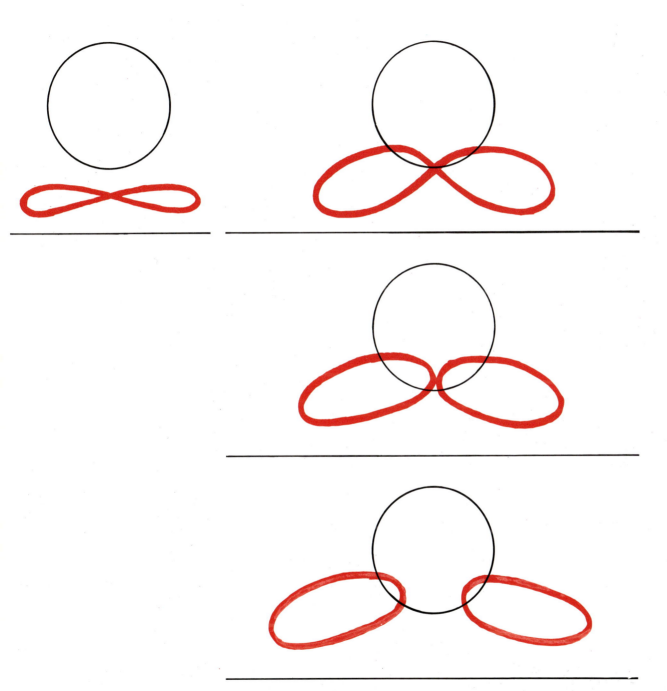

Ann and Betty are now both
moving round circles keeping a
fixed distance apart and Colin
continues to keep to the midpoint of this distance. This
motion might be created by
using a *three-bar linkage* as
shown. The two free ends may
be pinned to a drawing board
and a pen inserted in the middle
hole of the centre strip to trace
the locus.

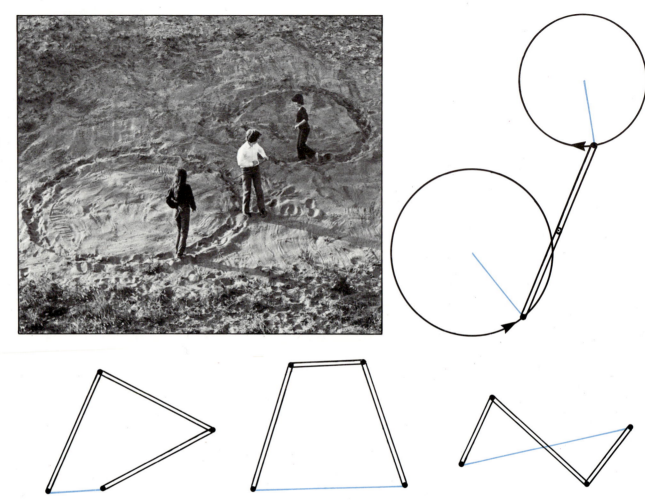

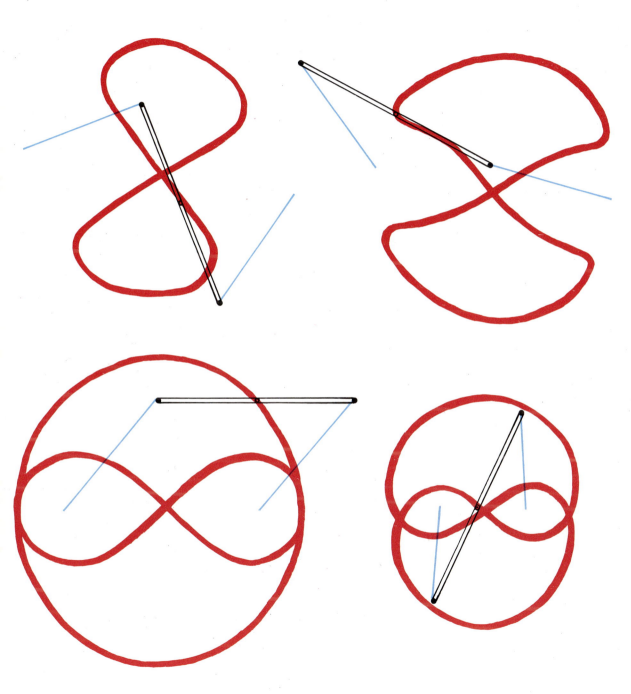

Various cases are shown for different distances between the fixed ends with the centre strip greater than the two equal side strips.

Here, there are again different
distances between the fixed ends
but this time the centre strip is
less than the two equal side
strips.

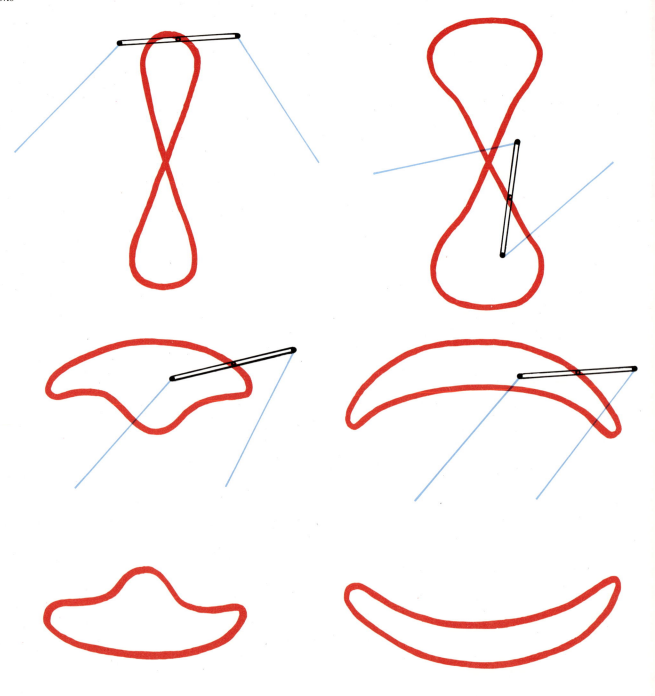

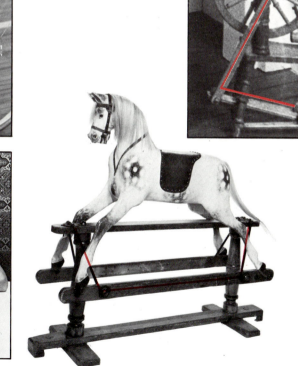

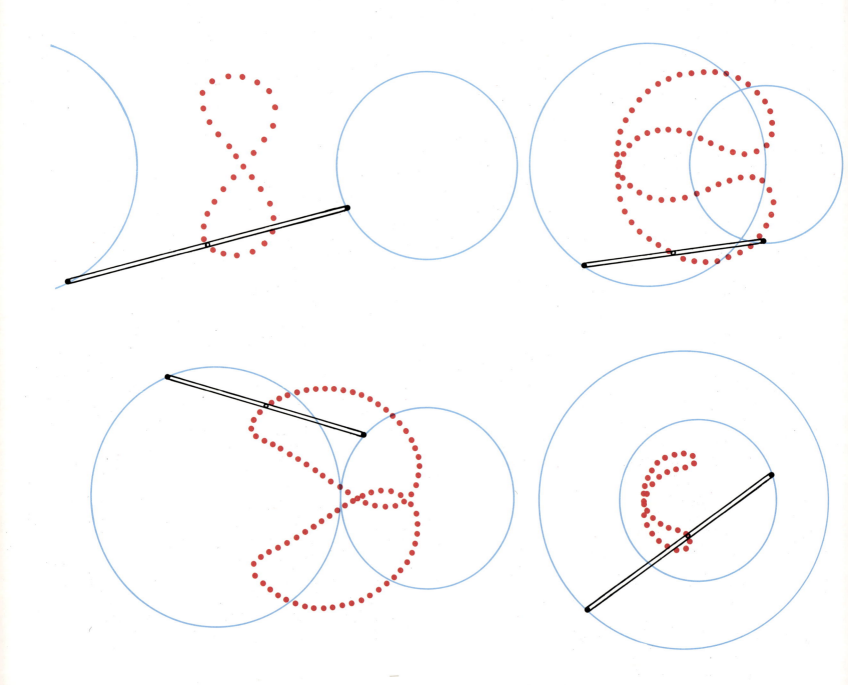

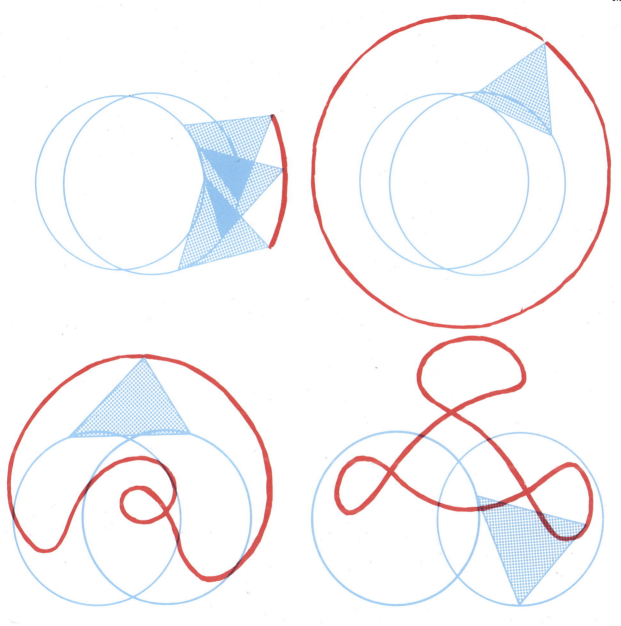

Generalised three-bar curves
determined by a point not on the
centre strip — here the locus is
generated by sliding a triangle so
that two of its vertices lie on
fixed circles.

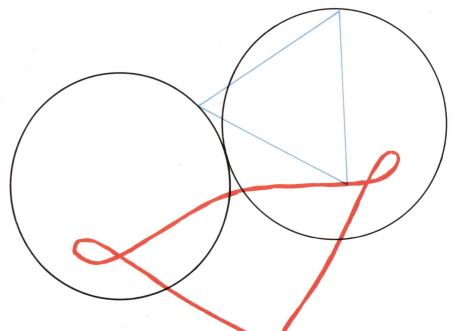

The parabola properties indicated on p43 also suggest another way of constructing a parabola as an envelope using a set-square. It is convenient to use a 45° set-square and to mark the perpendicular from the right angle vertex to the hypotenuse on the set-square. Alternatively, a shape of this short can be drawn and cut out from card. The perpendicular drawn on the set-square acts as a *guide-line*. The point where it meets the hypoteneuse is called a *foot-point*.

The parabola may be constructed by placing the guide-line always through a certain fixed point, with the foot-point always on a certain fixed line. For each position a line is drawn along the hypotenuse of the set-square. The diagrams show how this is equivalent to a construction of the parabola by paper-folding.

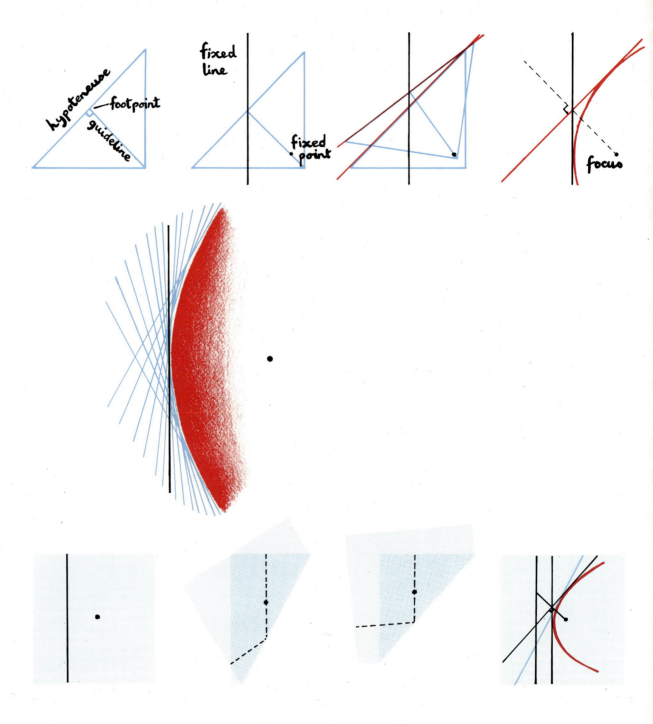

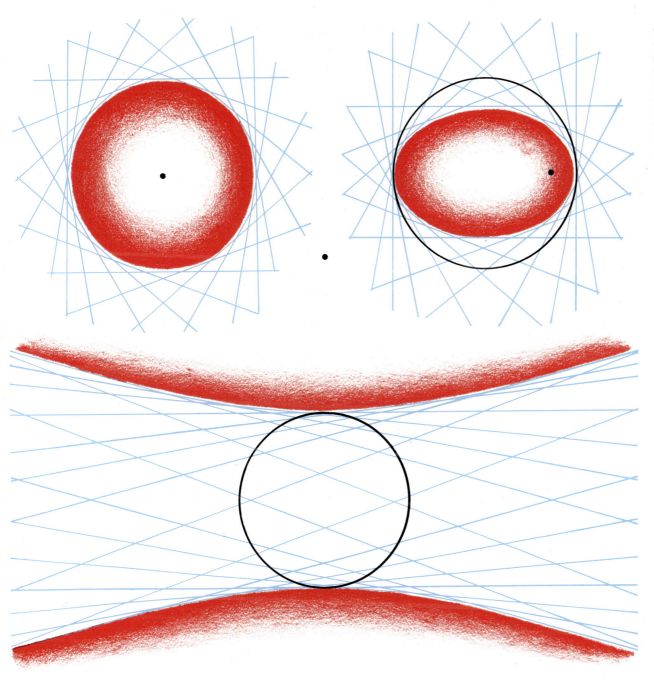

Other envelopes may be
constructed by placing the
guide-line as usual through a
certain fixed point and with the
foot-point always lying on a
fixed circle.
Different constructions are
shown for different positions of
the fixed point relative to the
fixed circle.

The set-square construction of the parabola shows that the foot-point lies at the foot of the perpendicular from a fixed point (in this case the focus of the parabola) to the tangents of the curve. The locus of such points is called the *pedal* (or *foot-curve*) of the original curve. Thus the pedal of a parabola with respect to its focus is a straight line. Conversely, the parabola is said to be the *negative pedal* of the line.

Set-square constructions can yield the pedals of a circle with respect to various points. These are *limaçons*, the particular case when the fixed point is on the circle being called a *cardiod*. Two of the drawings also show the pedal of a pedal...

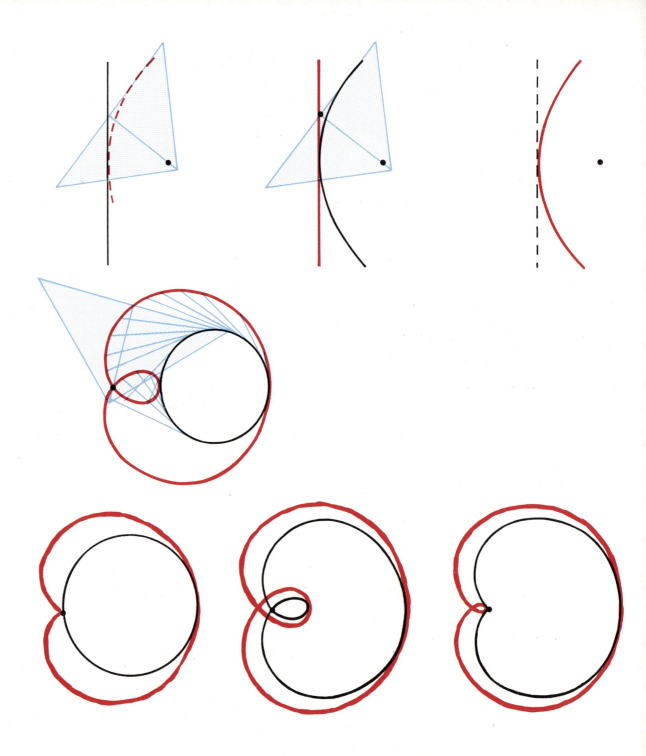

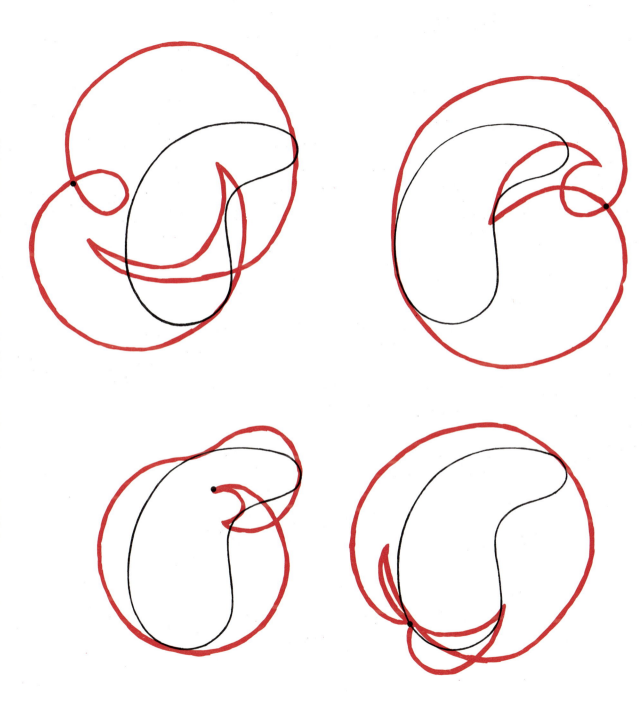

The set-square construction of pedals could be applied to any given curve. Here a random oval has been drawn and the pedal constructed for various positions of the fixed point.

Another set-square construction
is suggested in the diagrams.
Here the hypoteneuse of the
set-square always passes
through some fixed point while
the right-angle vertex always lies
on some fixed line. A curve is
traced by the various positions
of the foot-point.

Various cases are shown for
different positions of the fixed
point and the fixed line. When
these are the same distance
apart as the length of the guide-
line, the resulting curve is
known as a *right strophoid*.
Similar constructions where the
right-angle vertex now lies on a
circle, and finally on a freely-
drawn figure-of-eight, are shown
on the following pages.

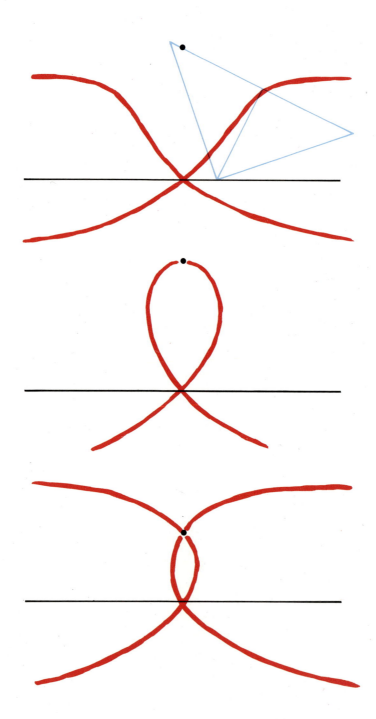

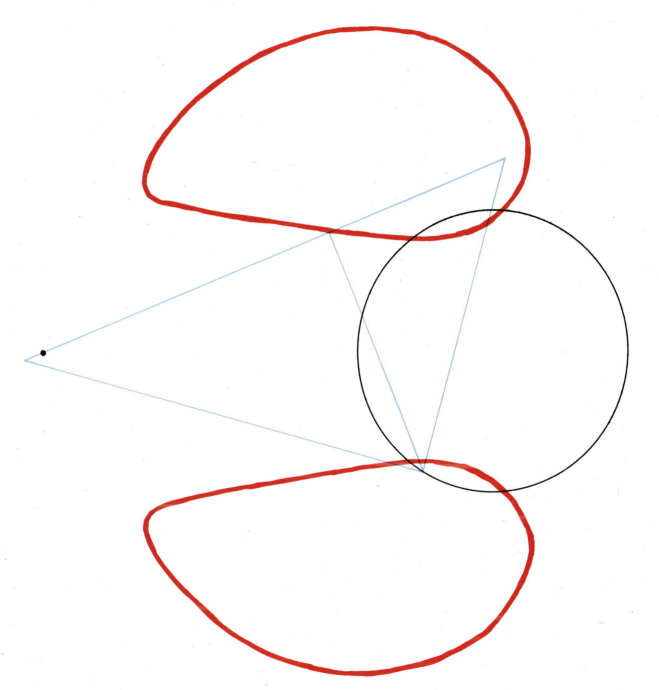

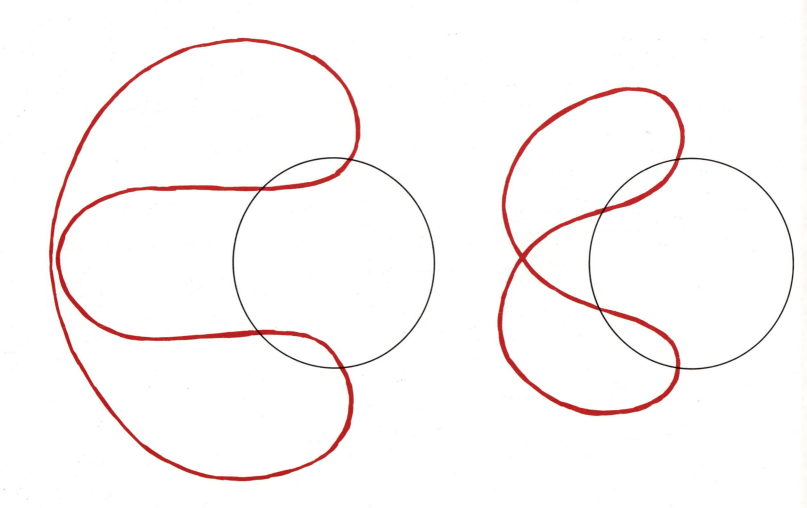

A set-square may also be used to construct certain spirals. This time the guide-line passes through a certain fixed point. The foot-point is placed at a certain chosen starting-point and then a new point marked at a fixed distance along the hypoteneuse from the foot-point. This is in turn chosen for the second position of the foot-point and so on.

When the fixed distance along the hypoteneuse from the foot-point is very small the resulting curve is very nearly an *equiangular spiral*. In this case the hypoteneuse lies along a tangent to the curve and the spiral is such that the angle between guide-line and tangent is always the same (in this case a right angle). This constant angle gives this spiral its name and the particular growth property that means it is found in so many natural forms.

As a variation the guide-line is made to pass through a moving point. Examples where this point moves round a circle during the construction are shown.

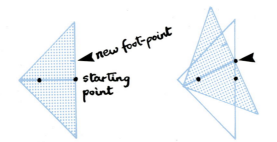

new foot-point

starting point

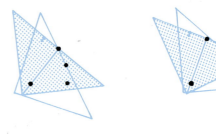

Glossary

Archimedes' spiral A spiral traced by a point that is receding at a constant rate as it turns. The curve is named after the Greek mathematician who first investigated its properties. It is referred to in the text as an equidistant spiral, since the whirls of the curve step out at equal intervals **p38**.

Astroid A four cusped hypocycloid constructed as a midpoint locus **p19** and a line-segment envelope **p55**.

Cardoid A one cusped epicycloid whose heart shape suggests its name. It can be constructed in various ways **pp19 24 48**.

Circle The locus of a point, or the envelope of a line, that moves in a plane keeping the same fixed distance from some fixed point. It can be considered as a special case of an ellipse but can also be seen as a special case of other curves. The simplest yet most primitively compelling of all curves.

Concentric A family of circles which share the same centre **p25**.

Conics The name given to the family of curves obtained by taking slices of a cone (hence their full title 'conic sections'). Depending on the direction of the slice these may be ellipses, parabolas or hyperbolas. Extensively studied by the Greeks, these curves may be constructed in many different ways. Conics are represented algebraically by quadratic equations. They are the most famous example of a branch of mathematics that was originally studied for its own sake but which later turned out to have important practical applications — especially as the orbits of planets and comets.

Cubics Curves that can be represented algebraically by equations of the third degree. They are not easy to construct by elementary means but some examples will be found on **pp 31 35**. Another class of cubic curves may be derived from the strophoid **p70**.

Curtate This word is sometimes used to describe the curves of a family that have bumps rather than loops, for example the middle curve on **p11** which is a curtate cycloid.

Curve-stitching A method of generating the lines that touch a curve. Stitching was introduced as an activity for young children by Mary Boole at the end of the nineteenth century. Interesting designs are often made by stitching parts of parabolas but other curves can be produced using similar methods **p42**.

Cusp A point on a curve where there is a sudden change of direction, so that the curve appears 'pointed'. In certain families of curves, the form with cusps can change to one with bumps or to one with loops **p10**.

Cycloid The curve traced by a point on the rim of a circle rolling on a line. This can be quite difficult to construct but the method offered on **p10** can be easily varied to yield a large class of related curves. The curve was extensively studied in the seventeenth century using the newly discovered methods of the calculus. The arch of a bridge is often cycloidal.

Deltoid A three cusped hypocycloid whose shape suggests the Greek letter *delta* and so its name **pp19 51**.

Directrix The name given to a certain fixed line associated with a conic. It is associated with a corresponding fixed point called the focus **pp43 66**.

Ellipse One of the family of conics; this may be thought of as a projection of a circle though various other definitions may be made. For example, it is the locus of a point that moves (in a plane) so that the sum of its distances from two fixed points is always the same; this gives rise to a pencil-and-string construction of the curve.

Envelope A set of lines determined by some constraints, for example the set of lines equidistant from a fixed point determines a circle, namely the one which all the lines touch. Any curve may be considered as being an envelope or a locus depending on whether the fundamental tracing element is thought of as being a line or a point.

Epicycloid Traditionally, the locus of a point on a circle that is rolling outside another fixed circle though another point of view is offered here **p18**.

Equiangular spiral A spiral traced by a point that is receding at a rate that is proportional to that with which it is turning. This describes a pattern of growth that is often found in natural forms such as shells, animal horns, fossils and flowers **p74**.

Equidistant spiral This name is used here in contrast to the equiangular spiral to describe the curve that is traditionally called Archimedes' spiral **p38**.

Focus Literally, a hearth or 'burning place'; the sun being at a focus of the earth's elliptic orbit. The bulb of a car headlight is placed at the focus of the parabolic reflector.

Folia A class of symmetric closed curves with intersecting loops **p17**.

Foot-curve A name for the locus of the point that is the foot of the perpendicular from a certain fixed point to a tangent to a curve. For example, an ellipse is the foot-curve of a circle from a point inside the circle. The foot-curve is often referred to in books as the pedal curve **p66**.

Foot-point, Guide-line, Guide-point These terms have been invented to describe a particular method of construction **p66** which uses a set-square on which a line (the *guide-line*) is drawn from the right-angle to the hypotenuse, meeting the latter at the *foot-point*. The constructions involve moving the set-square so that the guide-line goes through a point (*guide-point*) marked on the paper.

Hyperbola One of the family of conics, this curve may be thought of as an ellipse 'turned inside out'.

Hypocycloid Traditionally, the locus of a point on a circle that is rolling inside another fixed circle though another point of view is offered here **p18**.

Hypotenuse The side opposite the right angle in any right angled triangle.

Limaçon A single-looped epitrochoid named in the seventeenth century as being like the track of a snail **pp 15 24**.

Line Usually meaning a straight line which can be infinite in both directions. When infinite in only one direction it is sometimes called a half-line; when finite it is sometimes called a line segment.

Linkage An arrangement of line-segments pivoted together so that it can take up various positions, one, or more, points of the linkage being fixed. In practice, meccano strips loosely hinged, plastic geostrips or prepared strips of cardboard can be used. A circle can be considered as being drawn by a one-bar linkage. In principle any curve that can be represented by an algebraic equation can be constructed by some linkage.

Locus — A set of points determined by some constraints, for example the set of points equidistant from a fixed point determines a locus known as a circle.

Loop — The part of a curve that contains a double point, that is a point where the curve crosses itself **p11**.

Nephroid — A two cusped epicycloid whose kidney shape suggests its name **p19 48**.

Paper-folding — A method of constructing envelopes of lines by creasing paper **p66**.

Parabola — One of the family of conics, this may be thought of as occupying a boundary position between families of ellipses and hyperbolas, a sort of infinite version of either of these.

Pedal curve — See *foot-curve*.

Polar grid — A type of graph paper consisting of lines all passing through a point. Positions may be plotted by taking a specific distance from this point along a line usually indicated by a specific rotation from the horizontal.

Prolate — This word is sometimes used to describe the curves of a family that have loops rather than bumps, for example the last curve on **p11** which is a prolate cycloid (compare with *curtate*, above).

Quadratics — Name for an algebraic expression of the second degree and hence for the curves that can be represented by these (which are, in fact, all conics).

Quartics — Curves that can be represented algebraically by equations of the fourth degree. The limaçon is a traditional example.

Rose curves — A class of symmetric closed curves with many petals or loops. These arise as special midpoint constructions **p17** but can also be derived as foot-curves of epi- and hypo-cycloids.

Shear — A transformation in which each point of the plane moves parallel to a given fixed line. The distance moved by each point is proportional to its distance from the fixed line.

Spiral — The locus of a point rotating around and receding from a fixed point.

Star — A set of lines radiating from a fixed point.

Strophoid — A curve that is shaped like a *strophos* — the belt with a twist that carries a sword. This is a cubic curve whose construction **p70** can be interestingly varied and generalised.

Supplementary — Angles which together make up two right angles.

Tangent — Literally a line that touches a curve. In the case of a circle it lies on one side of the curve and meets it at one point. For other curves the situation can be more complicated, indeed the tangent can even cross the curve at the point where it 'touches'. The essential feature is that the tangent and the curve have the same gradient at this 'touching' point.

Three-bar and two-bar linkages — A set of 3/2 jointed rods which are free to turn about their joints in two dimensions.

Trammel — A device for constructing curves using pegs that slide in grooves. A typical example is the elliptic trammel invoked on **p54**.

Trochoid — A class of curves of which the cycloids are special cases. An epi- or hypo-trochoid is the locus of a point in the plane of a circle (but not necessarily on its circumference) that is rolling outside, or inside, a fixed circle. All the curves obtained by midpoint constructions are trochoids **p16**.

Viewing angle — A phrase coined to describe the angle subtended at the eye by a shape **p30**.